THE RIVERBANK

BY CHARLES DARWIN/FABIAN NEGRIN
CREATIVE EDITIONS
978-1-56846-207-3
PAGES: 32
PRICE(S): 17.95/25.65
GRADE LEVEL: 3 AND UP
PUB DATE: NOVEMBER 1, 2009

Annotation:

FOR FOSSILS AND STUDYING PLANTS AND ANIMALS, HE CAME TO BELIEVE THAT LIFE WAS EVERCHANGING AND THAT ALL LIFE FORMS ON EARTH COULD TRACE THEIR ORIGINS BACK TO A FEW PRIMITIVE ORGANISMS. THIS WONDERFUL PICTURE BOOK, CONCEIVED AND ILLUSTRATED BY ARTIST FABIAN NEGRIN, FEATURES WATERCOLOR PAINTINGS ACCENTED WITH PENCIL AND WAX THAT BEAUTIFULLY INTERPRET THE POWERFUL FINAL PARAGRAPH OF *ON THE ORIGIN OF SPECIES.*

Why we published:

THE RIVERBANK IS MOST SIGNIFICANT BECAUSE 2009 IS THE 150TH ANNIVERSARY OF THE PUBLICATION OF CHARLES DARWIN'S FAMOUS BOOK *ON THE ORGIN OF SPECIES. THE RIVERBANK* PRESENTS THE FINAL PARAGRAPH OF DARWIN'S BOOK, DEVOTING TWO PAGES AT THE END TO EXPLAINING KEY TERMS CONTAINED IN THE FEATURED PARAGRAPH.

IN *THE RIVERBANK*, A PICTURE BOOK CONCEIVED AND ILLUSTRATED BY ARTIST FABIAN NEGRIN, WATERCOLOR PAINTINGS ACCENTED WITH PENCIL AND WAX BEAUTIFULLY INTERPRET THIS STORY.

THE CREATIVE COMPANY

CREATIVE EDUCATION • CREATIVE EDITIONS • CREATIVE PAPERBACK

....

KIM LINGBECK

klingbeck@thecreativecompany.us

P.O. BOX 227, MANKATO, MN. 56002

507/388.6273 X227

PREFACE

When Charles Darwin first published *On the Origin of Species* 150 years ago, he shook science to its core. Not only did he outrage many by suggesting that the human species was descended from earlier animals, but he proposed that the arc of life is a never-ending contest he termed "natural selection." Every organism slowly changes, he argued, and nature decides if the change is beneficial and therefore preserved, or unbeneficial and therefore eliminated. Darwin offers his admiration of this gradual but grand progression in the following paragraph—the capstone of the book that rewrote natural history.

the Riverbank

Charles Darwin

illustrated by

Fabian Negrin

Creative Editions

Mankato, Minnesota

It is interesting

to contemplate

an entangled

bank,

clothed with
many plants of
many kinds, with
birds singing on
the bushes,

with various
insects flitting
about, and with
worms crawling
through the
damp earth,

and to reflect that these elaborately constructed forms, so different from each other, and dependent on each other in so complex a manner, have all been produced by laws acting around us.

These laws, taken in the largest sense, being Growth with Reproduction; Inheritance which is almost implied by reproduction;

Variability from the indirect and direct action of the external conditions of life, and from use and disuse;

a Ratio of
Increase so high
as to lead to a
Struggle for Life,
and as a conse-
quence to Natural
Selection,

entailing Divergence of Character and the Extinction of less-improved forms.

Thus, from the war of nature, from famine and death, the most exalted object which we are capable of conceiving, namely, the production of the higher animals, directly follows.

There is
grandeur in this
view of life,
with its several
powers, having
been originally
breathed into a
few forms or
into one;

and that, whilst
this planet has
gone cycling
on according to
the fixed law
of gravity,

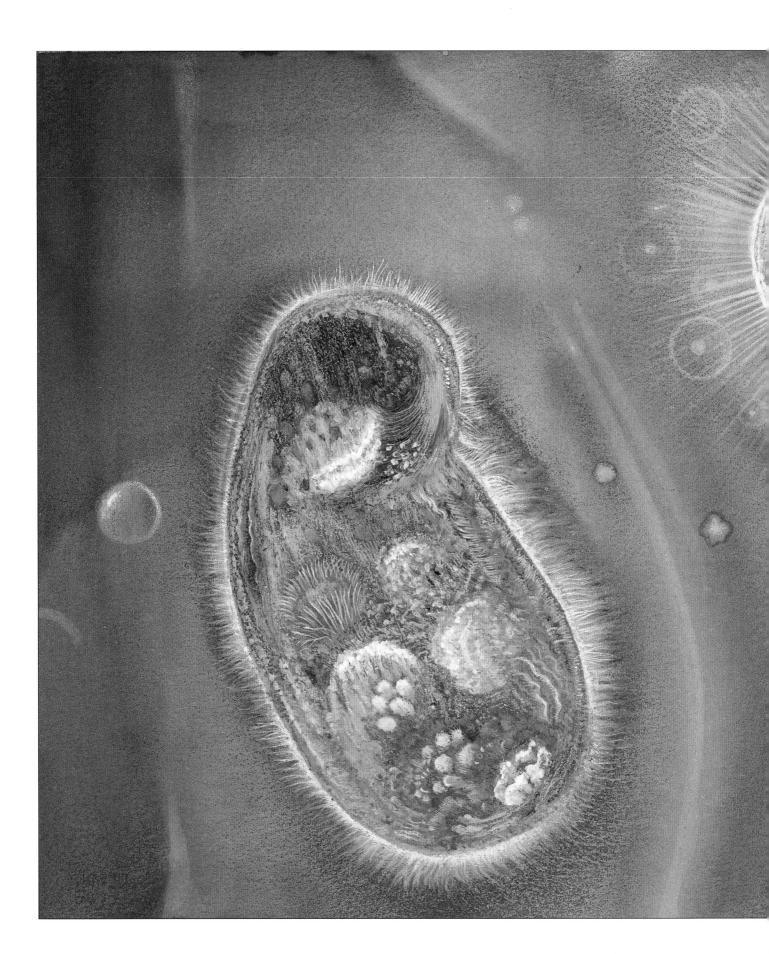

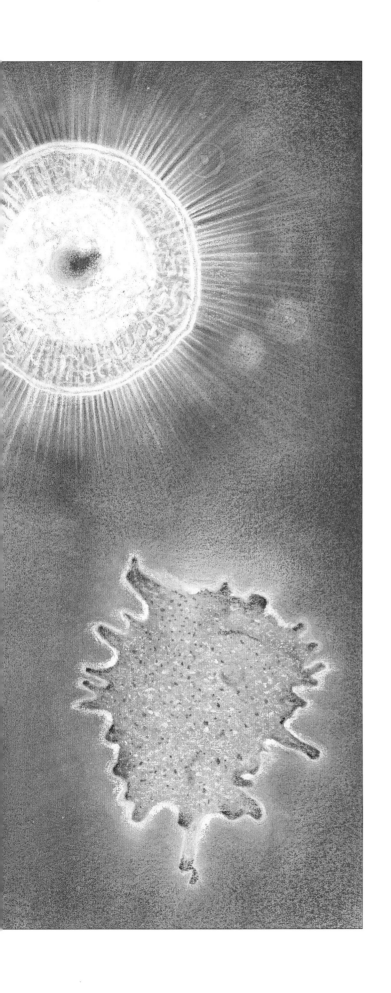

from so simple a beginning end-less forms most beautiful and most wonderful have been,

and are being,
evolved.

TERMS

laws basic rules that describe the workings of nature; natural laws are capitalized in Darwin's text

Growth with Reproduction the establishment of a species by the continued cycle of individual births

Inheritance the passing of traits from parents to offspring

Variability differences that arise among life forms of the same species because of where and how they live

Ratio of Increase a law that demands that, as any life form reproduces, a certain number of individuals must die so they do not overpopulate the environment

Struggle for Life the competition between life forms of the same species for food, living space, and mates

Natural Selection the process by which life forms slowly change, with some changes becoming permanent and others disappearing

Divergence of Character changes within one kind of life form until it branches into a different species

Extinction the permanent disappearance of a species

higher animals animals of intelligence, capable of learning

several powers a reference to the laws that describe the workings of nature

evolved changed over time in a way that better enables survival

For Giorgia, my love — F. N.

Illustrations copyright © 2009 Fabian Negrin

Published in 2009 by Creative Editions   P.O. Box 227, Mankato, MN 56002 USA

www.thecreativecompany.us

Creative Editions is an imprint of The Creative Company. Designed by Rita Marshall

Library of Congress Cataloging-in-Publication Data

Darwin, Charles, 1809-1882. The riverbank / text by Charles Darwin; illustrated by Fabian Negrin.

ISBN 978-1-56846-207-3

1. Evolution (Biology)—Juvenile literature. 2. Life—Origin—Juvenile literature.

I. Negrin, Fabian, ill. II. Title.

QH367.1.D37 2009   576.8—dc22   2008042674

First edition

2 4 6 8 9 7 5 3 1